Information Verification in the Digital Age

The News Library Perspective

Synthesis Lectures on Emerging Trends in Librarianship

Editor

Hema Ramachandran, *California State University, Long Beach,* and **Joe Murphy,** *Director Library Futures, Innovative Interfaces and Technology Trend Analyst.*

This series, *Emerging Trends in Librarianship* (a sub-series of the *Synthesis Lectures on Library Science and Librarianship*), will focus on new and emerging trends in digital collections and new technologies as they relate to the practice of librarianship and library science. The series will be of interest not only to librarians and information professionals, but also to the research community in general. Topics include but are not limited to: eScience, institutional repositories, data curation, advances in discovery tools, taxonomy and thesauri construction, mobile technologies, and the newest topic of "near field communication."

iv

Information Verification in the Digital Age: The News Library Perspective
Nora Martin

ISBN: 978-3-031-00911-2   print
ISBN: 978-3-031-02039-1 ebook

DOI 10.1007/978-3-031-02039-1

A Publication in the Springer series
SYNTHESIS LECTURES ON EMERGING TRENDS IN LIBRARIANSHIP #4
Series Editors: Hema Ramachandran, California State University, Long Beach, and Joe Murphy, Librarian & Technology Trend Analyst

Series ISSN 2372-8833 Print   2372-8868 Electronic

Information Verification in the Digital Age

The News Library Perspective

Nora Martin
NSW Ministry of Health, Sydney, Australia

SYNTHESIS LECTURES ON EMERGING TRENDS IN LIBRARIANSHIP #4

ABSTRACT

This book will contemplate the nature of our participatory digital media culture, the diversity of actors involved, and how the role of the news librarian has evolved—from information gatekeeper to knowledge networker, collaborating and facilitating content creation with print and broadcast media professionals. It will explore how information professionals assist in the newsroom, drawing on the author's experiential knowledge as an embedded research librarian in the media industry. The past decade has seen significant changes in the media landscape. Large media outlets have traditionally controlled news and information flows, with everyone obtaining news via these dominant channels. In the digital world, the nature of what constitutes news has changed in fundamental ways. Social media and technologies such as crowdsourcing now play a pivotal role in how broadcast media connects and engages with their audiences.

The book will focus on news reporting in the age of social media, examining the significance of verification and evaluating social media content from a journalistic and Information Science (IS) perspective. With such an emphasis on using social media for research, it is imperative to have mechanisms in place to make sure that information is authoritative before passing it on to a client as correct and accurate. Technology innovation and the 24/7 news cycle are driving forces compelling information professionals and journalists alike to adapt and learn new skills. The shift to tablets and smartphones for communication, news, and entertainment has dramatically changed the library and media landscape. Finally, we will consider automated journalism and examine future roles for news library professionals in the age of digital social media.

KEYWORDS

information verification, digital culture, embedded librarian, digital journalism, news sources, newspaper and media libraries, new media ecology, news reporting, social media

Contents

Acknowledgments

Special thanks to my editor Hema Ramachandran, who contacted me after reading my paper on Information Verification that I presented at the 2014 SLA Conference in Vancouver. Hema thoughtfully suggested that I build and expand on my existing research into information verification and provenance within the context of media librarianship for this series. My thanks also to Diane Cerra and the rest of the staff of Morgan and Claypool for their assistance and help along the way.

I gratefully acknowledge the expertise and generosity of the members of the News Division and Leadership and Management Division of the Special Libraries Association. I have been an active SLA member since 2007, and cannot speak highly enough about the value of belonging to this association. Special thanks to Libby Trudell, who kindly shared her article with me on crowdsourced content before it was published.

It would be remiss of me not to mention my lecturers from the Faculty of Arts and Social Sciences, University of Technology Sydney, who have made all the difference in my library career and academic pursuits, especially Dr. Michael Olsson and Dr. Hilary Yerbury. Their guidance, criticism and support over the years has been invaluable. I am particularly grateful to Maureen Henninger, Senior Lecturer for my Master's course *Investigative Research in the Digital Environment*. Maureen was able to provide me with insight into the recent emergent trend of data-driven journalism. Without her encouragement, I would not have been able to achieve much of the work I have described here.

Finally, I am indebted to David Babb, a New York-based engineering professional and scholar, who originally inspired me to pursue my postgraduate studies. Thank you for your wisdom and insight into the academic realm.

Dedication

This book is dedicated to my father, Dr. Kenneth Vincent Martin (1932–2008), who taught me the values of persistence and patience, instilling a hard work ethic. In 1957, Ken was awarded his Ph.D. in Organic Chemistry from the University of Sydney at just 24 years of age. My father subsequently undertook his post-doctoral research at the University of Illinois, and was employed as a Research Chemist based in Delaware. Our family moved to Australia in 1968. As a dual U.S./Australian citizen, it has always been a dream of mine to work in North America, so perhaps this tome may help to pave the way.

CHAPTER 1

Introduction

Newspaper and media libraries have existed since the advent of the print and broadcast media industry, but declining newspaper revenues in conjunction with technical advances have vastly changed the work of the news library professional in the digital age.

While the Internet can be a great resource for locating information, it is often difficult to know what trustworthy sites to consult in order to find the material that you are looking for. For example, many websites pronouncing themselves to be an authoritative source for federal government information actually have out-of-date or factually incorrect information listed. Finding those that are relevant and reliable can often be a challenging task.

Over the last decade, social media has become a fundamental part of our daily lives and professional activities. At the same time, there have been changes within the librarianship profession, such as the effects of social media on information sharing. This book explores a very critical aspect of the skill sets that citizens—especially "digital natives" need to develop in order to be credible and efficient communicators: the notion of information verification.

1.1   WHAT IS INFORMATION VERIFICATION?

News librarians, now usually referred to as news researchers (Barreau, 2005), are information professionals for whom information interaction is an essential part of their everyday work practices. Therefore the need to verify an individual's credibility, authenticity, and organizational affiliation applies to both librarians and journalists (Ojala, 2014). News researchers play a crucial role in the investigative process. Often they work collaboratively with print and broadcast media professionals to find authoritative information from secondary sources (Houston, 2009), such as an indexed clippings library, via digital repositories, or by performing routine and complex database searches.

Journalists need to check facts by validating and interpreting information. As one of the virtues of traditional journalism, verification is the resolve by reporters to examine evidence and test the veracity of any assertions that are made. In this context, interpreting often includes triangulating information from multiple sources (Blandford and Attfield, 2010; Shapiro et al., 2013). "Source triangulation" is a social scientific method that sets out to prove or disprove a hypothesis through triangulation of information from numerous sources and analysis of primary sources of official documents. Evidence-based practice is at the core of journalistic endeavours, with "information gathered according to rigorous principles and presented in the formats of conventional science" (Olsson, 2014: 81–82).

1.2   INFORMATION VERIFICATION AS A "CIRCULAR" PROCESS

Reveal[1] is an organization developing tools and services that aid in social media verification. As Reveal (2014) cites, one of the key findings of a pilot research study by Martin (2014) into the information practices of journalists is that verification becomes a circular process. In this study, participants agreed it was imperative to verify information and sources, noting that this was often a rigorous process until they were happy with the results (Martin, 2014):

> "… you verify what they tell you by cross checking and triple checking pieces of information, different sources that are unrelated preferably and that way you can cross check and verify and triangulate pieces of information, determine whether that piece of information is accurate or not…"

Due to the dynamic nature of the news gathering cycle, it may be necessary to relocate information that was initially discarded due to evolving requirements, or as new facts emerge. Consequently there is often an ongoing process of "identifying needs, finding and interpreting information that repeats before the information is used" (Blandford and Attfield 2010: 32–33).

While the filter is an apt metaphor for the model illustrating the steps in the news gathering process as displayed in Figure 1.1, there is an increased amount of information flow between the collection and verification/analysis stages due to the iterative nature of the news gathering process—this illustrates that information verification is a circular process.

---

[1]   http://revealproject.eu/about-reveal/.

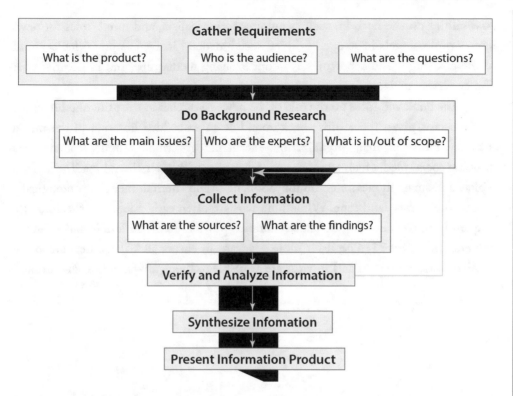

Figure 1.1: Steps in the news gathering process. Model based on Avramenko and Martin (2014).

1.3   SCOPE OF THIS BOOK'S DISCUSSION

This book is intended to give a comprehensive overview on issues of information verification[2] in the digital age within the context of media librarianship. Chapter 2 explores how information professionals assist in the newsroom, looking at the places that journalists often get "stuck" when doing their research. Building on this theme, Chapter 3 discusses news reporting in the age of social media, looking at how technology innovation and the 24/7 news cycle are driving forces compelling news researchers and reporters alike to adapt and learn new skills. Due to an increased emphasis on the principles of co-creation and crowdsourcing, Chapter 4 examines the significance of verification and evaluating social media content from an Information Science (IS) and journalistic viewpoint. This chapter highlights the links to the basic tenets of librarianship—critical evaluation of information. Chapter 5 provides a synopsis of

---

[2]   News-story verification is another term that is used to describe this process (Martin, 2015).

possibilities on the horizon, such as automated journalism, and future roles for news library professionals in an age of digital social media. Finally, Chapter 6 looks ahead to the challenges and opportunities in our evolving mediascape, and implications for library service delivery.

The book will focus on the role of the news researcher in the print and broadcast media sector, given the author's background as an embedded librarian in the media industry. However, many organizations—academic, government, and corporates, are utilizing social media tools to reach out and engage with their client base. The shift to tablets and smartphones for communication, news and entertainment has dramatically altered the library landscape. While librarians are often early adopters of technology, frequently our clients do not realize libraries provide access to authoritative online content. Thus, there is a need for more advocacy to ensure all stakeholders are aware of what library products and services are available to them, now and into the future.

CHAPTER   2

Information Professionals in the Newsroom

Libraries are very pertinent to their parent organizations and communities. Once considered the "gatekeepers" to knowledge, library staff in corporations and specialized settings have evolved from simply gathering information into the era of access,[3] delivering leading-edge information services to their clients (Ferguson, 2007; O'Connor, 2007).

In 1997, the British Library funded a preliminary one-year investigation called "Journalism and the Internet" in order to examine the changing information environment in the newsroom due to the Internet. Subsequently, *New Library World* published a feature article called "Journalists, News Librarians and the Internet." This study revealed that news librarians were much more positive about the internet than their journalist colleagues. At the same time, news librarians also had a heightened awareness of the problems, such as the authenticity or validity of retrieved information (Williams and Nicholas, 1997).

The authors concluded that while the internet was generally thought of as an end-user tool, it could potentially create novel and significant roles for news library professionals. It was predicted that librarians would work closer with journalists in the future, including possible involvement in more primary-sourced investigative work (Williams and Nicholas, 1997).

2.1   NEWSPAPER AND MEDIA LIBRARIES

The key objective of newspaper libraries is to support the reporting, editing, and illustrating of news stories by providing journalists with information and graphic materials, both in print and electronic forms. Most news libraries have collections of reference books,[4] journals (print and electronic), and an archive of published stories either in a clippings file or in an online database. Also, libraries offer their clients access to ex-

---

[3]   See http://libraries.pewinternet.org/2014/07/09/public-libraries-and-technology-from-houses-of-knowledge-to-houses-of-access/.

[4]   One essential reference guide of note is *Journalism: a Guide to the Literature*, by Jo A. Cates (2004).

ternal sources of information, regularly supplying material in a timely fashion to meet publication deadlines (Edds, 2003).

Magazines also employ news librarians, and in the broadcast media industry there are librarians on staff to assist with news broadcasts.[5] News librarians may be found in academic journalism libraries, and others work for vendors (Edds, 2003).

In the digital world, the nature of news has fundamentally changed. It has been reported that an industry wide crisis has triggered many news library closures (Paul and Hansen, 2002). In some cases, the closure of the library was due to the demise of the newspaper itself[6] (Murray, 2014). Declining newspaper revenue, as well as the self-sufficiency of reporters[7] are factors that have changed the nature of library work. This suggests the time is nigh for news library professionals to redefine their roles and reposition themselves within their organizations (Paul and Hansen, 2002; Matarazzo and Pearlstein, 2010). For instance, librarians need to be "champions for new technology," communicate value, and become leaders of technology initiatives. Moreover, they need to become more closely aligned to the business and develop partnerships, instead of just providing a service (Paul and Hansen, 2002; O'Connor, 2007).

In order to achieve such outcomes, some media organizations have assigned news librarians or researchers to editorial teams, raising their visibility among the journalists they serve (Barreau, 2005). This "embedded" structure means that librarians play an active role in the editorial process, which in turn facilitates the establishment of closer, more collaborative relationships with their library clients (Barreau, 2005; Brown and Leith, 2007).

There are five conditions for the sustainment and growth of embedded librarianship in corporate and specialized organizational settings (Shumaker 2012: 104):

1. Establish relationships with key decision makers at all levels.

2. Ensure that the quality and value of the embedded librarian's contribution continues to increase.

3. Lead the drive to perform necessary functions by the most cost-effective means available, whether in-house or by outside providers.

---

[5] For example, National Public Radio (NPR), CNN, and the Australian Broadcasting Corporation (ABC).

[6] For a list of news library staffing reductions from 2009 onward, see Quigley (2014).

[7] Examples of self-sufficiency include the ability of reporters to do their own research via the internet, and being able to access the digitized newspaper archive without having to consult library staff.

4. Adopt evaluation practices that are consistent with the parent organization's management culture.

5. In a financial crisis, seek to be part of the solution, not part of the problem.

The decision by Fairfax Media in Australia to decentralize their Research Library and embed information professionals within editorial teams created both challenges and opportunities for librarians.[8] As Brown and Leith (2007: 546) finds,

> *"Challenges include time management and prioritising in a highly deadline driven environment. Integration with the newsrooms and proximity to the journalists has led to a heightened sense of urgency with regards to the delivery of information .... Opportunities have also emerged .... A higher level of analysis of data by librarians is being required and carried out as the level of collaboration with the editorial team increases."*

Indeed, the close proximity between editorial staff and librarians increased the level of trust to such a degree that the Research Library team received an increased number of in-depth and complex requests. On many occasions, their timelines, summaries, and people and company profiles were published, unchanged, in Fairfax Publications; with the Fairfax Research Library credited with print acknowledgements. This raised the profile of the library both within and outside the company (Brown and Leith, 2007).

The practice of embedded librarianship is active, engaged, and customized (Shumaker 2012). It is evident that news librarians and researchers are knowledge networkers, shifting their focus from supporting content creation to creating content. While news library professionals still perform traditional tasks—such as compiling background information for a news report, they may also be proactive, taking the lead on primary research, or contribute original story ideas.

News librarians often act as trainers for newsroom staff, both in library applications and external database sources (Taylor and Parrish, 2009; Matarazzo and Pearlstein, 2010). For example, the Fairfax Media Information Services Department implemented advanced research training courses for journalists, run by their librarians. Sessions were held in the boardroom, so this meant it was convenient and accessible for journalists to participate. Training included: Advanced FDC (Fairfax Digital Collections, the internal database), as well as external database sources. Another idea under investigation was to offer training outside the head office premises: the concept of "roving librarians" traveling to all Fairfax community mastheads that have FDC, in

---

[8]   See Section 2.2: Case Study: Fairfax Media.

order to deliver training. This proposal would market and promote library services and facilities to journalists and staff at the same time.

To gain an understanding of the complexities of the role of the news librarian/news researcher, Matarazzo and Pearlstein (2010: 18) monitored the SLA News Division Newslib electronic discussion forum for a six-month period. Activities undertaken include:

- helping to find new revenue streams for their organizations;

- ascertaining how to use new technologies to enhance productivity and increase existing revenue sources:

  ○ digitizing text and photos (current and archival);

  ○ marketing network stock footage;

- responding to research inquiries from the public;

- using wikis as the backbone for an intranet;

- re-purposing blogs in print;

- utilizing social media sites as research resources;

- use of paraprofessional staff on the "help desk"; and

- investigating ways to reform copyright laws.

For library and information science professionals who aspire to work in the broadcast media sector, there are many areas of specialist knowledge required, extending from music to current affairs. You definitely need to be a news junkie and have excellent general knowledge. It is beneficial to be able to adapt what you know to unfamiliar territory. This means adapting your experience in traditional libraries to an audio-visual environment.

- Managing information: An understanding of digital libraries and content management, relevant metadata schemas, and production/client requirements is a prerequisite.

- Research: Apart from a sound understanding of content sources, it is crucial to have knowledge of copyright and restrictions in the broadcast environment. Much information and material retrieved for the media sector is necessary for re-use in production. This requires negotiation skills to clear copyright and obtain the rights needed by the production (One Umbrella Team, 2011).

2.1.1   JOURNALISTS' STUCK PLACES

While most journalists are happy doing interviews, many feel uncertain when it comes to delving into documents or databases (Houston, 2009). Yet documents and interviews can build on each other during the investigative process to produce a more in-depth news story, as displayed in Figure 2.1. A recent discussion on the SLA News Division Newslib listserv related to the places that journalists often get "stuck" when doing their research. In what ways do journalists typically have trouble finding, evaluating, and using information? Academic librarians are rethinking their approach, and are focusing on teaching to these "stuck" places.

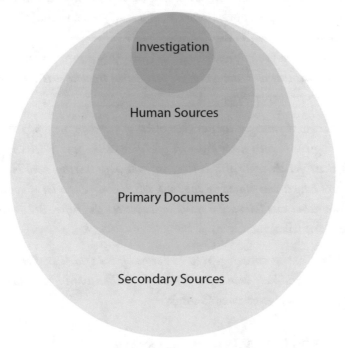

Figure 2.1: The investigative process: "Working from the outside in." Based on Houston (2009), p. 3.

Many reporters are intimidated by statistics, especially if they are the ones who say they hate math (and things like "this is why I went into journalism, so I wouldn't have to deal with numbers."). They don't understand why the most recent data available from X government agency is from two years ago. Or that if they call the agency, they might get a better answer for their question than what is found on the website. For example, the contact might be able to sort the data a different way or have reports not yet posted.

They also don't know how to check with associations, non-profits, and academics etc., as alternative sources for statistics. There may be some inherent biases in data from an industry association, but that doesn't eliminate them as a source if you take those into account. And they may be the only ones doing the counting (News Researcher, Texas).

[News Researcher, Texas] *raises a really good point about the value of seeking out alternative sources for statistics. And not just relying on a website but calling and asking for a data expert or a librarian. This need for alternative/diversity can also be extended to the sources used by reporters for interviewing. For example, here in Ohio, it's very easy to rely on Ohio State University experts for comment on stories. It's important that we reach out to get perspective from experts at other colleges and universities in the area, too.*

*I don't know if undergrads know the value of searching newspapers on microfilm and seeing the printed news coverage of a date in the past. With the Internet, you really don't get the sense of if a story is front page worthy, how big the headline was, and how it was played on the page. Microfilm gives a lot of good context. For example, what was the nation thinking about the day before 9/11? (*Archive and Collections Manager, Ohio).

I agree, we need to encourage reporters to go off the grid. Too often I find errors or stories that could have been better if the reporter had asked if we had a file (Online Producer/News Researcher, Georgia).

2.2   CASE STUDY: FAIRFAX MEDIA

For several decades the Fairfax Research Library provided a traditional service to staff, incorporating a central Information Desk. In response to calls to reduce the library footprint to accommodate 70 extra masthead staff, the Information Services Department decided to turn the situation into a positive, re-thinking how library services were delivered in line with international best practice. Hence, the Research Library was realigned from a traditional centralized services model, to an integrated team model (Brown and Leith, 2007).

In early 2006, I embarked on my corporate library career path as Research Librarian at Fairfax Media, one of the largest media organizations in the Australia Pacific region. My Communications degree combined with extensive work experience in an information service which had a client-service environment, including vast exposure to electronic databases, helped to secure the role. At this time, I worked closely with editorial staff and journalists in the newsroom and within a small team of information professionals responsible for reference and research work. Research Library staff were rostered around a seven-day week with hours from 9:00am to 11:00pm (Martin, 2007).

Just as technological changes affect information work, change also alters our physical setting. The Fairfax Research Library in Sydney was restructured just prior to my commencement (Brown and Leith, 2007). Instead of a traditional, centralized library, news researchers were moved onto the editorial floors, with a certain number attached to each masthead. This is a trend in other newspaper libraries, both in Australia and internationally. The *Age* Library in Melbourne has two staff situated on the editorial floor and *The Guardian* in the UK has a similar arrangement[9] (Martin, 2007).

The library's collections comprise a reference collection, a media and Australian history-focused book collection, and an online database of the full content of some 20 years of the Fairfax publications (in text and images), as well as a smaller range of PDFs (Brown and Leith, 2007). Library staff accessed a range of aggregated research tools, such as Factiva and the Nexis proprietary news database in the Lexis/Nexis product on behalf of clients, and managed a busy inter-library loan service. A separate Photo Library team facilitated access to electronic and hard copy images and provided a photographic reference service.[10]

Working in the world of current affairs and corporate media information is exciting, stimulating, and dynamic. My main area of focus was in the provision of reference services to journalists and non-editorial areas of the company. The extensive reference resources enabled the provision of diverse information to journalists such as the latest news on Kylie Minogue's social life, an explanation of Heisenberg's uncertainty theory, information on past winners of the Melbourne Cup, or a synopsis of Tosca. I was able to demonstrate proven customer services skills by reference to

---

[9] In 2008, *The Guardian* library team reverted to sitting together on a main editorial floor (Nelsson, 2012: 266).

[10] At a later date, the work undertaken by the Photo Library was outsourced to a commercial entity.

my ability to effectively deal with the research demands of journalists working in a pressured situation and to positively respond to changing work demands by reason of newspaper deadlines.

In this capacity, I conducted research for journalists on a daily basis, who have print deadlines to meet. When working on a Friday night which is the *Sydney Morning Herald* deadline, it was accepted practice to advise *Sun Herald* clients that priority must be given to the newspaper which is next going into production. This principle applied to all other deadline times. News library professionals must have the ability to determine priorities, meet strict deadlines and deliver the best possible client service at all times. Moreover, news librarians must be "equipped to meet the research needs of news reporters and editors who need information immediately" (Taylor and Parrish, 2009: 104).

The fast-paced, deadline-driven 24/7 news cycle brought fresh challenges on a daily basis. When I was attached to the *Sydney Morning Herald* masthead, I attended morning editorial conferences held by the *Sydney Morning Herald* journalism team. As a result, close working relationships developed between journalists and librarians, and the amount and intricacy of reference enquiries increased dramatically. I believe it is essential to be a proactive information provider, and often made suggestions to journalists regarding possible avenues of research after attending the morning conference.

Communities of Practice are increasingly important in the information profession. After noticing that there was no ALIA[11] e-list for librarians in the media sector, I devised an e-list specifically to create links between librarians in all areas of this sector, the outcome being improved professional standing of information workers in this sector through exchanging knowledge and experience. This initiative reaped rewards, with relationships being forged between Fairfax Research Library and the ACMA[12] Library as a result.

One of my duties was to assist with maintaining the daily updates of the electronic Fairfax archive, Fairfax Digital Collections (FDC). FDC is Fairfax's archival database that stores all published articles, photographs, and images. Part of my role was to maintain the text archive part of this database, classifying articles to assist in retrieval and searching. In addition, I assisted with maintaining the catalogue of the library's collection, updating the corporate intranet and our library intranet page, and administering loans of the hard copy files and index cards. In this context, I gained

---

[11] The Australian Library and Information Association (ALIA) is the national professional association for the Australian library and information services sector.

[12] The Australian Communications and Media Authority (ACMA) is responsible for the regulation of broadcasting, radiocommunications, telecommunications, and online content.

experience using the extensive Fairfax Media archive, comprising of card indexes and clipping files, enabling research on topics in the *Sydney Morning Herald* from early last century. The clipping files date from the 1930s to the end of 1995, and were consulted on a regular basis to assist with the research requests of journalists. New files were created to ensure ease in handling when they grew large.

2.3   CASE STUDY: AUSTRALIAN BROADCASTING CORPORATION (ABC)

In contrast to Fairfax Media, other organizations, such as the Australian Broadcasting Corporation (ABC), adopt a more traditional approach by maintaining centralized research services specifically a News Library, Sound and Reference Library, and Tape Library. As Australia's national broadcaster, the only programming requirements imposed on the ABC are general ones: the ABC must provide programs "of an educational nature"; it must broadcast regular sessions of news and information relating to current events within and outside Australia; and it must broadcast Parliamentary proceedings (ABC, 2005).

During 2013–14, I was employed as Research Librarian within the ABC Sound and Reference Library, providing high-quality research and information services to researchers, editors, reporters, and technologists across the organization. In this context, I was responsible for the provision of research support to ABC journalists and presenters from a wide variety of television and radio programs, such as ABC News, Lateline, 7:30 Report, Catalyst, Australian Story, Media Watch, Q&A, and Radio National. My responsibilities included delivering a specialized research service to journalists and staff operating in a deadline-driven environment.

Within the organizational structure, the Sound Libraries are part of the ABC Resources Division. Content Services manages the archives, libraries, rights and records service for the ABC. Within Content Services, the Sound Library network stretches across every State and Territory. Each library holds an extensive range of commercial and production music as well as sound effects. These recordings are available to ABC content makers in all locations throughout Australia.

The library's collections comprise reference collections, book collections, and serial titles focusing on broadcasting, arts, music, science and current affairs, and an extensive sound recording collection (CDs, production music, sound effects, and vinyl). Also, there are over 10,000 hardcopy newspaper clippings files. The libraries have specialist music librarians in each location, and offer direct client service to ABC staff each weekday from 8:30am to 6:00pm. Library catalogues are available on the ABC intranet, with reference and research across all collections.

Research librarians provide high-level research across news and current affairs databases with electronic document delivery provided by the Sydney and Melbourne libraries to all areas of the ABC, including international locations. It is essential for librarians working in the media space to maintain a thorough awareness of national and international news, current affairs, and politics to ensure comprehensive understanding of the broad range of client requests.

As part of the ABC Research Library team, my role involved contributing to the development and implementation of electronic resources, including review of existing services to improve availability and delivery options for clients. Sharing and collaboration of knowledge is vital in a fast-paced and dynamic broadcast media environment.

Collectively the ABC Sound Libraries hold approximately 230,000 CDs and 127,000 vinyl recordings. About 21,000 titles are added every year. There is a need to convert all existing analogue recordings to digital. This represents a huge challenge for the organization, given the volume of the existing CD and vinyl collection. Offsite sound transfers are managed through Radio Assist (Netia). Tracks can be transferred to any of 72 locations across the country. Over 8,000 tracks were transferred using Netia in 2013–14. The libraries currently have around 3,000 active members and loan approximately 90,000 items per year.

The ABC Sound Libraries maintain a service agreement with the State Orchestras for assistance with their programming and planning requirements, allowing access to all orchestra members but loans only to administrative staff.

The ABC is covered by blanket rights agreements with the following music industry organizations:

- Australasian Mechanical Copyright Owners Society Limited (AMCOS)

- Australasian Performing Right Association Limited (APRA)

- Australian Record Industry Association Ltd (ARIA)

- Phonographic Performance Company of Australia Limited (PPCA)

Rights vary according to usage; live to air, pre-recorded or online podcasts, and a number of other restrictions exist. The libraries service a number of radio and television networks.

CHAPTER 3

News Reporting in the Age of Social Media

In the digital world, the nature of news has fundamentally changed. The balance of power between news media and the audience has been altered, with a power shift in the digital age from "journalist as gatekeeper" to the citizen as editor (Kovach and Rosenstiel, 2010). Bloggers and user-generated content are inextricably woven into the news production process, the result being an incorporation of varied content, diversification of source material, and multiplicity of actors. In other words, there has been a democratization of the news gathering process, with the differentiation of journalism between professionals and citizens. As Jarvis (2007) explains, "witnesses to events can now help report what they see and context and explanation can come from both journalists and the experts they quoted who can now also publish....I see that not as a competitive threat, but as a grand opportunity."

The speed of change in our connected world shows no sign of slowing, with mobile platforms the most disruptive force for news content delivery and audience consumption. A report by the Pew Research Center (2016) analyzing the landscape of American journalism highlights the growing influence of technology, with 62% of adults reporting that they use social media as their main news source. Similarly, a 2016 Australian survey reveals that for 52.2% of respondents social media was their main source of news, and the top source of news among online media was through websites or apps of newspapers (21.7%) (Watkins et al., 2016).

3.1 NEWS CONTENT CREATION

Social media and technologies such as crowdsourcing play a pivotal role in how broadcast media connects and engages with their audiences—interactive information sharing. There have been corresponding changes in journalism norms and practices, with the rise of the citizen journalist. Howe (2006) defines crowdsourcing as "the act of taking a job traditionally performed by a designated agent (usually an employee) and outsourcing it to an undefined, generally large group of people in the form of an open call."

As an exemplar, the ABC (Australia) newsgathering structure means that the organization ensures that social media is taken into account during the planning, newsgathering, and production phases. The ABC news team are using their Instagram account to showcase photos by staff reporters and citizen journalists. Professional journalists are engaging with the Instagram community, liking and commenting on their photos, and requesting to use photos in news multimedia coverage. This "crowdsourced" approach has already borne fruit, with Instagrammers alerting ABC journalists to news photos by "tagging" the photos in question. In effect, anyone with a smartphone can become an extension of the newsgathering operation (Posetti, 2009). Moreover, social media searches allow reporters to follow their own breaking news on social media platforms, to ascertain audience reaction and follow-up if required (Gearing, 2014).

Jarvis (2006) defines this participative approach as "networked journalism," noting that "it takes into account the collaborative nature of journalism now: professionals and amateurs working together to get the real story, linking to each other across brands and old boundaries to share facts, questions, answers, ideas, perspectives. It recognizes the complex relationships that will make news. And it focuses on the process more than the product." Hence, "newspapers" are yielding to "news content" as the news environment and news formats themselves become increasingly complex and virtual (Cheney et al., 2011).

News content is more observational; citizen media, Twitter, and blogs are generating news as it happens (Cheney et al., 2011; Hermida, 2012). News events and announcements are often replayed on Twitter well before they appear on news websites (Jericho, 2012). On the day of the G20 protests, journalists from *The Guardian* in England and commentators supplied a constant stream of live Twitter updates. Eyewitness accounts were published, unmediated by editors, via a feed on *The Guardian* website, along with maps, and a live blog. The numerous layers of coverage meant users received an extraordinarily rich sense of what occurred as events unfolded (Ahmad, 2010). Clearly "crowdsourcing" can produce riveting personal tales as well as a wide range of expert opinion, which often results in new story possibilities being unearthed and explored (Kovach and Rosenstiel, 2010).

3.2   NEWS CONTENT DISTRIBUTION

The news industry has been redefined, evolving to exploit newer media. New distribution models have been created, with consumers no longer relying on just a few major established news companies. Increasingly, news is being generated through hyperlocal news sites and models such as ProPublica[13] (Cheney et al., 2011). In general, there is less original reported news and a greater number of news aggregators.[14] This blurs the distinction between the news creators and those who provide access to the news. For this reason, many consumers will not be able to discern the difference between a news aggregator and a news agency (Cheney et al., 2011). As such, librarians need to be aware that this may present challenges when assisting clients in locating and using news content for research.

Due to our 24/7 news cycle, news is more immediate, with users at the heart of content delivery. News content is delivered based on one's personal preferences and interests, and preferred delivery model (Cheney et al., 2011; Hermida, 2012). Users can create their own content and upload it to social media platforms, controlling what content they want to view and when they want to view it.[15] It is de rigueur for news to be conveyed by way of a personalized news stream, filtered by a social network of friends, instead of via traditional media (Hermida, 2012).

Changes in delivery and access include paywalls/pay per view systems being established, as news publishers and providers of all sizes devise ways to "monetize" their content (Cheney et al., 2011). The digital subscription model was introduced by *The Wall Street Journal* in 1995, paving the way for other well-known titles such as *The New York Times*, *The Times of London*, and *The Australian* (Bennett, 2015a). In 2008, Rupert Murdoch addressed this issue, announcing that News Corporation would charge for accessing the digital version of *The Wall Street Journal* by providing three tiers of content: "The first will be the news that we put online for free. The second will be available for those who subscribe to wsj.com. And the third will be a premium service, designed to give its customers the ability to customize high-end financial news and analysis from around the world" (Murdoch, 2008). By 2010, *The Wall Street Journal* had 414,000 paid digital subscribers. In 2015, the *WSJ* digital subscriber count is more than 900,000, with combined digital and print sales in excess of 2.2 million (Bennett, 2015a).

Another case in point is *The New York Times*, which signed up 100,000 subscribers in the first three weeks after instituting a paywall. Subscribers were charged from $15 to $35 per month, with non-subscribers being able to access limited web content (Bates, 2011). By 2015, *The New York Times* had over one million paid digital

---

[13] ProPublica is an independent, non-profit newsroom that produces investigative journalism in the public interest, https://www.propublica.org/.

[14] For example, Google News, *Huffington Post* and Slate.com.

[15] Some examples are podcasts of radio programs and catch-up television services via free-to-air networks.

subscribers (Bennett, 2015a). As Herford notes, "*New York Times* sets the agenda not only for a lot of [Washington] DC, but also for discussion among the leadership class. Same is true for the *Wall Street Journal* and *Financial Times* in business" (Herford 2012, quoted in Hooke, 2013). Undoubtedly, the subscription model works for high-end titles that target affluent and educated consumers[16] who are willing to pay for quality news content.

In recent times there has been a rethink on digital subscription models, with growing support for free models starting to emerge. It appears that often audience reach can be limited by subscription models, with readers needing to be convinced that paid-for content is superior to content that is freely available via the web. For instance, popular UK daily *The Sun* curtailed their two-year-old subscription model in late 2015. Similarly, the *Toronto Star* in Canada abandoned their digital model in April 2015 after less than two years (Bennett, 2015a). As Earnshaw (quoted in Bennett, 2015a) observes, "the industry needs to find a model that works, especially for the younger consumers that have grown up in a world of free ubiquitous content and who have never bought a newspaper. This generation expects content for free and with that mindset there is no going back." Nevertheless, while Millennials may stumble across news via the Internet, this behavior is not a model for sustained future news engagement, or an educated democratic society (Poindexter, 2015).

3.3   IMPACT OF NEWS CONTENT TRENDS ON LIBRARIES

Digital news has forever altered the print news industry, creating complex challenges for librarians and researchers. Information professionals need to stay abreast of trends in the news industry, and be fluent in the use of traditional news databases,[17] news aggregators,[18] and non-traditional news sources when searching for news online (Sabelhaus and Cawley, 2013). In the future, free resources such as Google News may reduce the use of traditional databases. Consequently, technology innovation and the 24/7 news cycle are driving forces compelling information professionals to adapt and learn new skills. Knowledge of advanced Boolean search strategies, combined with

---

[16] Also known as the AB socio-economic group. The Socio-Economic Status (ABC Scale) was established by the National Readership Survey as a form of classifying and describing social classes for media outlets.

[17] Traditional news databases include Dialog, the Nexis proprietary news database in the Lexis/Nexis product, and Factiva.

[18] News aggregators include Google News, HighBeam Research, and PressDisplay.com.

using complex search features available via news aggregators are vital skills for library professionals (Sabelhaus and Cawley, 2013).

Key questions about how digital content can be preserved and archived have only recently been asked, four decades after newspapers began storing news digitally.[19] From the worldwide Millennium celebrations of 2000, to the Paris attacks of 2015, big news stories now break first online, not in print. Stories are updated in real-time, and they continually shift and change as new information comes to light. They are viewed on mobile devices, and shared via social media channels. Yet their online presence is short-lived, with news headlines visible and viable only for a distinct time period (Skinner and Graham, 2015). As McCain (2015a: 338) points out: "All born-digital content has proven to be ephemeral in the extreme: susceptible to bit rot, technical failure, human error and natural disaster." According to the Pew Research Centre's Project for the Excellence in Journalism (2012), already "the emerging world of community online news, less than a decade old, can be difficult to access." Once a single print object that could be microfilmed, news is now a series of digital streams. Publishers gather various feeds from sources and add it to their own reportage. Hence, news is not only converging, it is also a composite. This represents cultural norms and original forms of reporting, thus creating issues and complexities with regard to the archiving of news content (refer to Figure 3.1).

Preserving born-digital content requires an active and ongoing set of activities designed to keep the content viable even as computer hardware, operating systems, applications, file formats, and other aspects of the technology platform constantly evolve (McCain, 2015c). The importance of preserving born-digital content is obvious, given that the consequences can be dire should a mishap occur. In 2002 the *Columbia Missourian* server hosting its out-of-date content management system crashed, and the sole system back-up system in place failed. Approximately 15 years of news articles and 7 years of photos were lost—forever (McCargar, 2006). To prevent such a scenario in the future, the Reynolds Journalism Institute is a building born-digital photo archive at the *Columbia Missourian* which will act as a test laboratory for journalism (McCain, 2015c).

---

[19] Special Issue: "Capturing and Preserving the 'First Draft of History' in the Digital Environment," *Newspaper Research Journal* 2015, Vol. 36 (3), covers a wide range of subjects re: archiving of news content in a digital age.

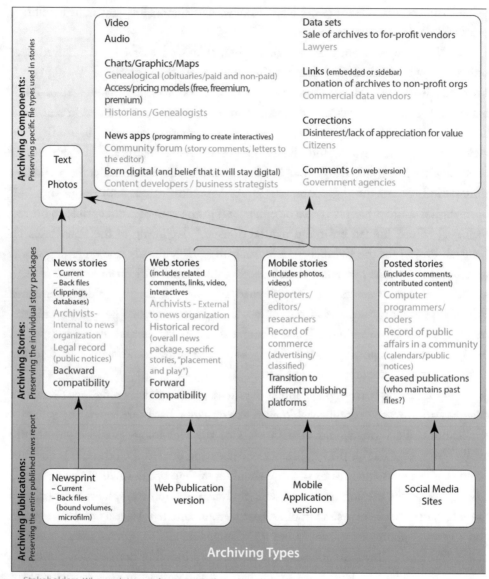

Figure 3.1: Archiving types. Based on Hansen and Paul (2015a, p. 286).

3.4   HOW SHOULD LIBRARIES RESPOND?

Cheney et al. (2011) argue that libraries need to challenge their assumptions relating to the provision of access to and preserving news content. A 2012 survey of news librarians found that most U.S. newspapers are maintaining their digital news files primarily for near-term access (five years or less), and are not yet seeking to ensure the longevity of this record.[20] This means that librarians need to explore innovative approaches that facilitate the preservation of born-digital news content. Yet the question of how libraries of today can begin the task of preserving our digital news content into perpetuity continues to be a hot topic of debate.

A 2014 survey of news publishers by the Reynolds Journalism Institute in Missouri revealed that most U.S. media enterprises fail to adequately process their born-digital news content for long-term survival.[21] Edward McCain is the digital curator of Journalism at the University of Missouri Libraries and leads the Journalism Digital News Archive (JDNA), which addresses issues surrounding access and preservation of digital news collections. In an email to the author, McCain (2015b) maintains that: "This is a pretty huge topic, of course, and I think there is no one answer to the question of how libraries should respond. Right now it may well be that we need to be responding in a number of ways in order to discover the more successful approaches."

A survey by Hansen and Paul (2015b) from the University of Minnesota examined the archiving practices in ten newspapers. Nine of these news organizations were legacy newspapers, and one was digitally native. Their findings emphasize the complexity and variety of news assets that need to be preserved, as well as the organizational challenges in newsrooms where different systems, software and staff are responsible for the creation and control of these assets. The traditional, legacy system of storing newspaper archives in folders, bound volumes, or on microfilm was unsatisfactory. Content stored this way eventually deteriorates, or can get lost or misfiled, particularly when news organizations close, merge, or change ownership. Concurrently, digital storage seems to be just as problematic: none of the news organizations surveyed have clear policies for archiving and preserving online resources (Hansen and Paul, 2015b).

---

[20] Educopia Institute/The Bishoff Group. 2012 Survey on Long-Term Archiving of Born Digital Newspaper Content.

[21] McCain, Edward. Reynolds Journalism Institute and University of Missouri Library. 2014 Survey on Newspaper Preservation Practices.

Curating data for long-term access requires that content be actively prepared in order to be preserved (Skinner and Graham, 2015). Chronicling America (http://chroniclingamerica.loc.gov/) is one example of digital content curation. This is an online database of historic U.S. newspapers published from 1836–1922. The site makes digitized newspapers available through the National Digitized Newspaper Program, a joint program of the NEH and the Library of Congress. The NDNP data is public domain, with the site being published as open-source software to facilitate the implementation and customization by other organizations keen to digitize their own newspaper collections. Technical aspects are based on sustainable practices in digital preservation, thus ensuring content will be accessible in the future. This includes adopting open and standardized file formats and metadata structures, technical validation, and using digital collection and inventory management tools developed by the Library of Congress (Engle, 2015).

Citizens worldwide depend on historical news sources for a multitude of purposes. Librarians and archivists perform a stewardship function, helping to preserve cultural heritage, one part of which resides in newspaper collections. While discussions on the topic of digital news preservation are underway between key stakeholders,[22] more conversation and work needs to be encouraged across business sectors (including cross-disciplinary collaboration) to address this crucial issue, and help save the "first rough draft of history."

---

[22] E.g., "Dodging the Memory Hole" is an ongoing series of outreach events by JDNA designed to facilitate cooperation, collaboration, and alignment among media companies, memory institutions, and key stakeholders.

CHAPTER 4

Evaluating Social Media Content

The exponential growth of social media as a central communication practice, and its agility in capturing and broadcasting breaking news events more rapidly than traditional media, has forever altered the journalistic landscape: social media has been adopted as a significant source by professional journalists, and conversely, citizens are able to use social media as a form of direct reportage. This creates new opportunities for newsrooms and journalists by providing an avenue for newsgathering via access to a wealth of citizen reportage and updates about current affairs, as well as an additional showcase for news dissemination. This chapter discusses verification tools and strategies to further understand and to make the most of social media content for library and information science (LIS) practitioners. Aggregation of news has become simultaneously easier with technology, yet more challenging due to the abundance of news sources. As a result, there are "major challenges with vetting the quality of news sources" (Cheney et al., 2011: 74).

4.1 VETTING THE QUALITY OF NEWS SOURCES

> "'Citizen Journalism' and 'crowdsourcing' are new trends that require the same probing, questioning, and analysis before a journalist makes use of them in a story."
> Kee Malesky on the topic of vetting information sources, in Altschiller (2011).

The European Journalism Centre, with partners that include the Poynter Institute for Media Studies,[23] asked a team of media analysts at the cutting edge of verification to compile a book on how best to use user-generated content (UGC). The free *Verification Handbook* (http://verificationhandbook.com/) is a valuable guide to the verification of digital content, offering theoretical and practical tips on how to closely inspect videos, images, and other UGC, outlining case studies which illustrate the strategy used in different situations. Craig Silverman is a renowned fact-checking

---

[23] The Poynter Institute for Media Studies is a major contributor to discussions on trending issues in the media, http://www.poynter.org/. In June 2015, the First Draft Coalition (with the help of Google News Lab) launched a new site, "First DRAFT News: your guide to navigating eyewitness media, from discovery to verification," http://firstdraftnews.com/about/ (Hare, 2015).

champion, editor of *The Verification Handbook*, and author of the Regret the Error blog (http://poynter.org/tag/regret-the-error/). Silverman (2012) contends that "Never has it been so easy to expose an error, check a fact, crowdsource and bring technology to bear in service of verification."

One verification strategy is checking a Twitter user's profile for a blue check-mark badge, which means that Twitter has verified that user's identity (Wardle, 2014; Phillips, 2015). However, often there is no quick way of verifying the identity of a social media account. Instead, this verification of source requires "painstaking checks," including reviewing the source's network and previous social media postings. Preferably, the journalist should speak directly with the original source of the information (Wardle, 2014).

Learning how to cut through the confusion intrinsically present in social media is critical and makes verification "the essence of journalism" argues Steve Buttry (2014), who discusses how the ways of accurately verifying news have evolved over time, adapting to the new technologies. These new technologies have contributed positively to news creation while making it easier to manipulate content, or to create content to mislead—as propaganda for misinformation purposes. As a result, "the ease of digital video editing—like that of other digital techniques, raises the importance of scepticism" (Buttry, 2014). In a similar vein, Malachy Browne (2014) from Storyful[24] explains how he fact-checked some video footage, by wisely cross-referencing digital tools such as Google Street View and Twitter. The video was uploaded onto YouTube by an athlete competing in the Boston marathon, depicting the explosion along the route.

One of the key points about the use of social media in a crisis is that first, it should do no harm. Talking about the humanitarian crisis in the Nigerian city of Jos, Stèphanie Durand (2014) insists that "social media perpetuate misinformation, while at the same time enabling journalists to connect and interact with members of the public as part of their work." Furthermore, Durand (2014) finds "social media also provide a platform to respond to rumors, and verify information that ultimately creates the type of trust and transparency necessary to avoid an escalation of conflict." Often, the best response of Twitter users not directly involved in a tragedy is silence, so as not to create confusion or risk smothering important and useful information with noise. A case in point, the tweeting of cat photos during the Brussels lockdown in 2015 (Rogers, 2015).

---

[24] Storyful specializes in finding and verifying content on social media. So if you are a client of Storyful, you can email them and they will run a check on the content.

Mathew Ingram (2014) writes about the importance of knowing how to create a network of reliable sources on Twitter, a method he calls "responsible crowdsourcing." A good example is the work of Andy Carvin from NPR and his "sifting through the social media streams" of sources out of the Middle East in 2011 (Hermida, 2012: 321). Not only does Carvin follow them on Twitter, he has formed trusted relationships, which in turn means that verified information can be acquired. For instance, a photograph of a mortar which exploded in Libya was sent to Carvin and broadcast to his network of Twitter contacts. Carvin's sources helped him to identify the type of weapon and its country of production, establishing that the bomb was not of Israeli origin (Ingram, 2014).

The final chapter lists various digital tools that can be used to fact-check UGC content (http://verificationhandbook.com/book/chapter10.php). The list is divided into three distinct categories: verification of the identity of the sources; localization of the content; and verification of the truthfulness of images, and also includes a list of other useful tools.[25] Some key points: Reverse image search allows you to quickly check to see if an image has been posted online before. If you use Chrome or Firefox you can install add-ons that make it as easy as right-clicking on the image to do the search. Above all, the Golden Rule is to get on the phone whoever posted the content (BBC, 2016).[26]

4.2   CREDIBILITY OF ONLINE INFORMATION

The University of Maryland LibGuide JOUR479M (http://lib.guides.umd.edu/JOUR479M) is a useful resource for librarians who would like to learn more about understanding social media use in journalism. This resource outlines a number of criteria that can assist in locating credible sources of information (University of Maryland Libraries, 2016). These are often grouped into the following categories:

- authority/accuracy;

- purpose and content;

- currency; and

---

[25] See also "Tools for verifying and assessing the validity of social media and user-generated content," http://journalistsresource.org/tip-sheets/reporting/tools-verify-assess-validity-social-media-user-generated-content.

[26] See "BBC processes for verifying social media content" for various case studies that illustrate the verification process, http://www.bbc.co.uk/blogs/collegeofjournalism/entries/9117dbc6-db68-30fd-9787-41c7da51a85c.

- design/organization/ease of use.

Criteria: Authority/Accuracy

When judging the authority of a web resource, contemplate the following:

- Author

  ○ Who is providing the information?

  ○ What are their qualifications?

- Affiliation

- URL indicators (.edu, .gov, .net, .org)

- Links—where does the site lead you?

- Contact information.

One method to ascertain who owns a website/domain name is to perform a WHOIS search to find a domain registry service. This can provide an indication as to its authority. (http://whois.net/) (http://www.networksolutions.com/whois/index.jsp).

When judging the accuracy of a web resource, consider such questions as follows:

- Is it untidy or full of errors?

- Are sources provided?

- Has the site been evaluated?

- How did you find this site?

  ○ Broad search engine?

  ○ Selective directory?

  ○ Link from a reputable source?

Criteria: Purpose/Content

When assessing the purpose and content of a web resource, think about the following:

- Stated or implicit purpose

- ○ Advocacy

- ○ Commercial use

- ○ Hoax/Fabrication

- ○ Informative

- ○ Humor/Satire

- ○ Personal page

- ○ Propaganda

- ○ Hacked information

- • Coverage

  - ○ What topics are included

  - ○ In depth exploration of topics

- • Evidence of bias

  - ○ Is there a minimum of bias?

  - ○ To what extent is information featured on the site of a persuasive nature?

Criteria: Currency

Depending on the purpose of the web resource, currency may also need to be taken into account. Reflect on the following:

- • Update frequency: Is this historical data, or does it need to be current?

- • Currency of links: Are links from the site current, or are they out-of-date?

Criteria: Design, Organization, and Ease of Use

These factors help to determine the usability of a page. Consider questions such as:

- • Does the layout serve the user?

  - ○ Is it logical and well-organized?

  ° Does it contain "help" features?

- Is the design logical and easy to follow?

- How much scrolling is required to find the needed information?

- Are buttons and boxes large enough?

4.3   LINKS TO BASIC TENETS OF LIBRARIANSHIP

"Journalists, like librarians, are trained to evaluate sources, to question the authority, currency, and comprehensiveness of any piece of information or statement of fact." Kee Malesky on the links between journalistic practices and librarianship, in Altschiller (2011).

In 2011, *Library Journal* published an interview with Kee Malesky, National Public Radio's eminent research librarian.[27] In this piece, Malesky reveals valuable insights about NPR reference techniques, her views on the paradox of the Internet, and the demands of working in a deadline-driven environment (Altschiller, 2011). Instead of a centralized library service, NPR librarians are embedded throughout the newsroom and within various desks and shows. As Sanders (2013) notes, "Even from the beginning, research and librarians were so important to NPR that we hired an information specialist to assist staff with story ideas and background information even before we hired the reporters."

Library staff receive over 11,000 reference queries per year, produce briefing books for major events, work on investigative projects, as well as maintain an internal wiki (Sueiro Bal, 2012). The NPR librarians perform research duties for the entire organization, including the News department. Library staff are expert at searching the web, accessing major commercial databases such as the Nexis proprietary news database in the Lexis/Nexis product, and utilize search engines such as Google Advanced on a regular basis. As not everything is available in full-text on the Internet, local libraries and a document-retrieval service are often used to obtain resources. In addition, librarians routinely contact primary sources on the phone or by email,[28] thus ensuring accuracy and timeliness (Altschiller, 2011).

---

[27] Kee Malesky is often referred to as NPR's longest-serving reference librarian, http://www.npr.org/2010/10/23/130729448/all-facts-considered-by-nprs-longtime-librarian.

[28] Primary sources include: government agencies, academic experts, etc.

Working as a research librarian in a deadline-driven environment can be challenging.[29] If library staff are only given a short timeframe to answer a question, they may offer alternative information instead. At all times, NPR librarians diligently exhaust every resource and reach out to any expert to satisfy the journalist's need for information. For questions requiring extensive historical detail, librarians use online and print sources, compiling data so that it can be sorted and analyzed, confer with outside experts, and deliver a more thorough report than standard requests would require (Altschiller, 2011).

Professionally, Malesky always meticulously cites sources, and believes that the Internet has added to the propagation of unsourced and questionable facts: "It's really the paradox of the Internet: that so much information and primary material is available instantly and anywhere, but careful thought and analysis are required to ensure that the information is correct, current, authoritative, and therefore useable" (quoted in Altschiller, 2011).

In April 2011, the *Beyond Books*[30] symposium sponsored by Journalism That Matters,[31] the MIT Center for Future Civic Media, the American Library Association, the Media Giraffe Project, and the New England News Forum brought together like-minded librarians, journalists, and citizen activists. Joy Mayer, Associate Professor at the Missouri School of Journalism focuses on community engagement and how news and information can be more of a conversation. Reflecting on the Beyond Books event, Mayer (2011) suggests that librarians and journalists have many values in common, such as:

- the belief that actionable knowledge is necessary for an empowered democracy;

- the belief that critical thinking skills are of fundamental importance;

- a desire to elevate the quality and diversity of community discourse; and

- a role to play in helping people find quality information.

With such shared values, librarians and journalists are therefore positioned within their communities as trusted sources of information and are committed to responsibly informing citizens. This notion is summed up eloquently by NPR librarian Hannah Sommers (2013), "One idea that defines us is... a belief as professionals that

---

[29] For an in-depth look at NPR library, http://www.npr.org/sections/npr-extra/2012/08/23/159910472/a-look-to-the-future-one-archive-tape-at-a-time.

[30] Beyond Books, http://journalismthatmatters.org/biblionews/.

[31] Journalism That Matters, http://www.journalismthatmatters.net/.

everyone should have access to quality information to make the decisions that life and work require."

CHAPTER 5

Future Possibilities

There is no doubt that the workplace is changing. Information professionals are faced with new technologies, new generations of users and evolving user needs. In the midst of this change, however, there are opportunities to continue to demonstrate the relevance of the news library specialist. This chapter will present an overview of the changing landscape for library and information science (LIS) practitioners. It will look specifically at the opportunity presented by big data and the influence of automation in the workforce, and review practical examples of how media organizations are demonstrating value in this space.

5.1 AUTOMATED JOURNALISM

People often think of robots as mere science fiction, yet we come across them every day. Siri makes suggestions on your smartphone; an intelligent personal assistant Alexa is built into a dedicated speaker called Amazon Echo; self-driving cars are on the horizon; minimally invasive surgery is performed using robotic technologies such as the da Vinci System. Similarly, with the rise of automation in journalism, automated tools such as algorithmic software can be used to write texts from structured information (Wang, 2016).

A topic of much debate, the automation of news content is now underway at prominent media companies including the Associated Press, *Forbes*, *Los Angeles Times*, *The New York Times*, and ProPublica (Graefe, 2016). When an earthquake happens in Los Angeles, an algorithm publishes an article on the *Los Angeles Times* website.[32] Another algorithm at the masthead writes about homicides (http://homicide.latimes.com/). The Associated Press also uses an algorithmic platform called Wordsmith, developed by tech company Automated Insights that produces news reports that are able to be "personalized and stylized to match a company's editorial voice" (Mumford, 2014). This means that Associated Press could then generate approximately 4,400 earnings report stories per quarter, compared to the 300 it was once delivering with its own reporters (Mumford, 2014). Due to their data-driven nature, such news stories

---

[32] The *Los Angeles Times* was the first newspaper to publish a story about an earthquake using algorithmic software on March 17, 2014, http://www.latimes.com/local/lanow/earthquake-27-quake-strikes-near-westwood-california-rdivor-story.html.

are ripe for automation. It is interesting to note that Automated Insights CEO Robbie Allen believes that while most journalists want to write one article that will be read by many people, his company's goal is quite different: "we'll create a million pieces of content that we hope a million people read just one of" (quoted in Rutkin, 2014).

A study by Christer Clerwall (2014) of Karlstad University in Sweden placed articles on the Internet, one written by a journalist, the other a software program. Participants were asked to rate how they perceived each article using a select amount of key terms. Readers rated the automated article as more descriptive, more accurate, trustworthy, and objective. The journalist's article was rated more interesting, pleasant to read, well written, and less boring. However, when asked who had written the article, a robot or a human journalist, many participants could not discern the difference. While such experiments ignore crucial factors that readers consider when assessing news in a real-world context, they show that "there is at least a place for automated journalism to grow and develop" (Shanley, 2015).

Recent research in this area by Columbia University's Tow Center for Digital Journalism has outlined key findings for news consumers, as follows (Graefe, 2016):

- People rate automated news as more credible than human-written news but do not particularly enjoy reading automated content.

- Automated news chiefly suits topics where providing facts in an expedient way is more important than refined narration, or where news did not exist previously and consumers therefore have low expectations about the quality of the writing.

- There is little knowledge about news consumers' need for algorithmic transparency, i.e., whether they need (or want) to understand how algorithms work.

Many scholars and practitioners see the technology's potential to enhance news quality; however, others are more cautious—fearful that automated journalism will eventually equate to job losses (Graefe, 2016). Yet AAP[33] editor-in-chief Tony Gilles insists, "This isn't about replacing human journalists; it's about giving us more of an opportunity to take more news on, with the same number of staff" (quoted in Mumford, 2014).

Dr. Andreas Graefe, of the Ludwig Maximilian University of Munich is engaged in a number of studies to ascertain how readers react to computer-generated content. In an interview with Nicola Holzapfel (2016), Graefe argues that with rou-

---

[33] AAP—Australian Associated Press.

tine tasks being assigned to algorithms, journalists have more time for sense-making, "We will see robots producing first drafts, which the journalists will then edit and supplement. I spoke to a sports reporter at AP about this, and he told me that he used to have to begin writing his report as soon as the game was over. Now as algorithms can summarize the game, he actually has time to interview the players."

Although Graefe concedes that job losses were inevitable for reporters principally focused on routine tasks, new roles are also emerging. As an exemplar, the Associated Press now employs an automation editor who is tasked with finding new possibilities for automation. Furthermore, in order for new algorithms to be created, trained journalists are required who can "stipulate how a news report should be structured and can recognize the relative importance of news items" (Holzapfel, 2016).

5.2   THE EVOLVING ROLE OF THE INFORMATION PROFESSIONAL

There is a paradigm shift occurring in the library and information sector. Leveraging data and evaluating research portfolios are common services being offered. No longer simply gatherers and managers of data, librarians increasingly provide analyses and interpretations of the data. Often social media specialists, they also require a well-rounded repertoire, such as multimedia content production skills, storytelling and communications skills, data analytics, and critical analysis capabilities (McCosker et al., 2016). Information professionals skilled in these tools and techniques are poised to play a critical role in their organization.

Robyn Shulman (2016) has offered a characterization of how the librarian's role continues to evolve: "librarians are not only caretakers of books and information tour guides; they are learners, researchers, teachers, and curators. In addition, today, most librarians are digital media and technology specialists. They manage books, curate content, provide research guidance and structured pathways for all who come through their doors. In addition, they are continuously learning, teaching their patrons, and maintaining the entire digital information literacy ecosystem."

In order to examine the emerging roles and possible futures for librarians and information professionals, Vassilakaki and Moniarou-Papaconstantinou (2015) undertook a systematic literature review of the specific roles librarians and information professionals have adopted in the past 14 years. Even though a range of libraries were included, the authors note that "the majority of the literature focused on academic libraries." Their findings have implications due to the impact of technology on our everyday work practices, combined with changes in education and society, and how those changes affect our institutions. Six emergent roles were discovered: librarian as teacher, technology specialist, embedded librarian, information consultant, knowledge manager, and subject librarian. All of these roles appear to be relevant in a variety of library workplace contexts.

In a chapter on News Libraries in *British Librarianship and Information Work 2006–2010*, Richard Nelsson from *The Guardian* states that from 2002 onward new roles that media librarians undertook included building information architectures, managing e-content information, and working side-by-side with their journalist colleagues (2012: 265). *Guardian* librarians were credited in the paper and online, developed a research intranet and a blog, and the team "continued to adapt, moving into areas such as data-journalism" (2012: 266).

News librarians are adaption masters, leading the way in innovation in terms of "information architecture, site navigation, database construction and analysis, taxonomy, and digital archiving" (Montgomery, 2010). In today's data-driven newsroom, lines are even more blurred. Incrementally building on the "embedded" researcher theme, often there are no longer distinctions made between research library and editorial staff (Berry, 2015). Thus, librarians and journalists alike must embrace technology, be agile, and adapt to change in our disruptive news milieu (Slaughter, 2011; Nelsson, 2014; Holt, 2016).

CHAPTER 6

Conclusions

In the digital era, news librarians have moved beyond the reference desk, becoming embedded news researchers and knowledge networkers, collaborating and facilitating content creation with print and broadcast media professionals. Data journalism is frequently the result of investigative teams, with the news researcher now one of many co-creators.

In 2005, veteran journalist Bill Kovach presented a keynote speech at the *Society of Professional Journalists Convention*. In his address, Kovach (2006) stated that "if journalism of verification is to survive in the new Information Age then it must become a force in empowering citizens to shape their own communities based on verified information."

As Kellie Riordan (2014), 2014 fellow at the Reuters Institute of Journalism, points out, "the verification of information, especially in the fast-paced viral news world, remains the greatest challenge of the digital news revolution." Most importantly, there is a strong correlation here with one of the basic tenets of librarianship—critical evaluation of information. In a world saturated with user-generated content, it is imperative that citizens—especially "digital natives"—have the skills to critically evaluate sources of information.

6.1 THE CHANGE IN NEWS: CHALLENGES AND OPPORTUNITIES

On LinkedIn Pulse, Paul Shanley (2014) from the Associated Press discusses the five trends that will reshape the news industry in the next three to five years; his list is as follows:

- the growth of video;

- migration to mobile devices;

- continued migration of consumer attention to social media;

- the data-driven newsroom; and

- automated journalism.

One of the keynote speakers at the 2015 *Future Forum: Influencing a Connected World* in Sydney, Raju Narisetti, Senior Vice President of Strategy at News Corp, discussed the challenges faced by the news industry, noting that their resolution requires a fresh examination of current practices. In an era of information and co-creation, newsrooms should be "gate-openers, not gate-keepers." Narisetti (2015, quoted in Cheng, 2015a) argues that "while our core strength is still the journalism and information we create, our greater opportunity is combining that with the trust readers put in our ability to help them navigate an ever enlarging sea of information and news."

The data-driven newsroom must focus on audience preferences, and deliver a contextualized user experience. Indeed, the push to personalize the news involves paying attention to user behavior (Colhoun, 2015). In relation to reader experience vs. content, Narisetti believes that publishers should combine their print, digital, and IT teams to ensure they create vibrant news experiences that will mean audiences keep returning (Bennett, 2015b).

Another speaker at the 2015 *Future Forum: Influencing a Connected World* in Sydney was Lauri Baker, Vice President, Brand Strategy and Sales, Head of Branded Content at *Huffington Post*. Baker (2015, quoted in Cheng, 2015b) said that publishers need to recognize that readers consume content differently across social networks, as "social is the new front page." The social media explosion is driven by an increase in user-generated content and changing consumer preferences. The *Huffington Post* built their own content management system that linked with Google Analytics and search engine optimization. In brief, the company reinvented the way that publishing works. Baker (2015, quoted in Cheng, 2015b) explains it this way: "when we were publishing content, we knew in real time what people were searching for… and we could change that, and optimize that, every five minutes, every day, two days later, a week later … so every day those stories were relevant."

6.2   IMPLICATION FOR LIBRARY SERVICE DELIVERY

At this stage, it is uncertain what role user-generated content will play in relation to library service delivery (Trudell, 2014a). An informal poll of library and information professionals by Trudell identified interest in the topic. Yet, most poll respondents reported minimal use of crowdsourced content. In an email to the author, Trudell (2014b) concludes: "Not surprisingly, very few information professionals are currently using crowdsourced materials in any regular way, but I think that will evolve."

Service providers offering credible crowdsourced content exist, with more expected to materialize in the future. Key questions to ask if you are considering incorporating such resources into your research portfolio include (Trudell, 2014a):

- Is the organization offering the content or service known and credible?

- Does the service maintain and demonstrate a clear editorial or quality control process?

- Does the service engage contributors with expertise in the field or topic?

- Does the organization have a sustainable business model?

Crowdsourcing is being used to create content such as product reviews, feedback, and ideas (Staff of the Worthington Library, 2010). While the role played by crowdsourcing for enterprise information is emergent, libraries should be ready to take advantage of this opportunity. Library clients will expect to participate interactively, and their input should be encouraged (Staff of the Worthington Library, 2010).

As the ways clients seek, access, and use information changes, library staff must adapt by responding intelligently to evolving user requirements. Given the impending emphasis on using social media for research purposes, it is imperative to have mechanisms in place to make sure that information is authoritative before passing it on to a library client as correct and accurate.

Bibliography

Ahmad, A.N. (2010). "Is Twitter a useful tool for journalists?" *Journal of Media Practice* 11 (2): 145-155. DOI: 10.1386/jmpr.11.2.145_1. 16

Altschiller, D. (2011). Q & A: Kee Malesky, NPR Librarian. *Library Journal* 136(7): 118. 23, 28, 29

Australian Broadcasting Corporation (ABC) (2005). *ABC Fact Sheet: The ABC's Charter, Independence and Accountability.* Australian Broadcasting Authority, pp. 1–2. 13

Avramenko, O. and Martin, N. (2014). Assessment task 2: Presentation of a case study investigation. *57152: Investigative Research in the Digital Environment. Master of Arts in Information and Knowledge Management*, University of Technology, Sydney. 3

Baker, L. (2015). Day 2: Plenary - Mobile Future, paper presented to the *2015 Future Forum: Influencing a Connected World.* Sydney, Australia, September 10–11.

Barreau, D. (2005). Integration of information professionals in the newsroom: Two organizational models for research services. *Library and Information Science Research* 27: 325–345. DOI: 10.1016/j.lisr.2005.04.012. 1, 6

Bates, M. E. (2011). Interacting with the news. *Online.* September/October: 64. 17

BBC Academy. (2016). User-Generated Content and the UCG Hub. BBC. Retrieved February 9, 2016, from BBC: http://www.bbc.co.uk/academy/journalism/skills/social-media/article/art20150922112641140. 25

Bennett, L. (2015a). To Pay, or Not to Pay. *The Bulletin.* November, p. 11. Retrieved January 15, 2016, from NewsMediaWorks: http://www.newsmediaworks.com.au/the-bulletin-is-out-now-2/. 17, 18

Bennett, L. (2015b). Video: Narisetti on reader experience vs content. *NewsMediaWorks.* October 29. Retrieved May 28, 2016, from: http://www.newsmediaworks.com.au/video-reader-experience-rules-over-content-quality-narisetti/. 36

Berry, C. (2015). Personal conversation with the author, October 19, 2015. 34

Blandford, A. and Attfield, S. (2010). Interacting with Information. *Synthesis Lectures on Human-Centered Informatics*, Lecture #6. Morgan and Claypool. DOI: 10.2200/S00227ED1V01Y200911HCI006. 1, 2

Brown, D. and Leith, D. (2007). Integration of the research library service into the editorial process: "Embedding" the librarian into the media. *Aslib Proceedings: New Information Perspectives* 59(6):539–549. DOI: 10.1108/00012530710839614. 6, 7, 11

Browne, M. (2014). Verifying Video. In: *Verification Handbook: A Definitive Guide to Verifying Digital Content for Emergency Coverage*, edited by Craig Silverman. Maastricht, the Netherlands: European Journalism Centre. Retrieved February 9, 2016, from http://verificationhandbook.com/book/chapter5.php. 24

Buttry, S. (2014). Verification Fundamentals: Rules to Live By. In: *Verification Handbook: A Definitive Guide to Verifying Digital Content for Emergency Coverage*, edited by Craig Silverman. Maastricht, the Netherlands: European Journalism Centre. Retrieved February 9, 2016, from http://verificationhandbook.com/book/chapter2.php. 24

Cates, J. A. (2004). *Journalism: a Guide to the Reference Literature*. 3rd edition, Libraries Unlimited. Westport, CT. 5

Cheney, D., Palsho, C., Cowan, C., and Zarndt, F. (2011). The Future of Online Newspapers. *Proceedings of the Charleston Conference*. DOI: 10.5703/128824314878. 16, 17, 21, 23

Cheng, A. (2015a). Newsrooms should be 'gate-openers, not gate-keepers.' *NewsMediaWorks*. July 29. Retrieved May 28, 2016, from http://www.newsmediaworks.com.au/newsrooms-should-be-gate-openers-not-gate-keepers/. 36

Cheng, A. (2015b). Video: Huff Po's Lauri Baker on social as the new front page. *NewsMediaWorks*. October 22. Retrieved May 28, 2016, from http://www.newsmediaworks.com.au/video-social-is-the-new-front-page-lauri-baker/. 36

Clerwall, C. (2014). Enter the Robot Journalist. *Journalism Practice* 8(5): 519-531. DOI: 10.1080/17512786.2014.883116. 32

Colhoun, D. (2015). Is the news behaving more like advertising? *Columbia Journalism Review*. May 1. Retrieved April 11, 2016, from http://cjr.bz/1FCU3Pw. 36

Durand, S. (2014). Case Study 1.1: Separating Rumor from Fact in a Nigerian Conflict Zone. In: *Verification Handbook: A Definitive Guide to Verifying Digital*

Content for Emergency Coverage, edited by Craig Silverman. Maastricht, the Netherlands: European Journalism Centre. Retrieved February 9, 2016, from http://verificationhandbook.com/book/chapter1.1.php. 24

Edds, C. (2003). Frequently Asked Questions about Newspaper Libraries, Retrieved October 27, 2015, from SLA News Division Website: http://www.ibiblio. org/slanews/about/faq.htm. 6

Engle, E. (2015). Extra! Extra! Chronicling America Posts its 10 Millionth Historic Newspaper Page. October 7. Retrieved January 29, 2016, from Library of Congress Digital Preservation Blog: http://blogs.loc.gov/digitalpreserva- tion/2015/10/extra-extra-chronicling-america-posts-its-10-millionth-his- toric-newspaper-page/. 22

Ferguson, S. (2007). Introduction. In: *Libraries in the Twenty-First Century: Chart- ing New Directions in Information Services*, Stuart Ferguson (ed), Topics in Australiasian Library and Information Studies, Number 27. Wagga Wagga, NSW: Centre for Information Studies, Charles Sturt University. DOI: 10.1108/01435120810917620. 5

Gearing, A. (2014). Investigative Journalism in a Socially Networked World. *Pacific Journalism Review* 20(1): 61–75. 16

Graefe, A. (2016). Guide to Automated Journalism. *Tow Center for Digital Journalism*, Columbia Journalism School, New York City. Retrieved April 18, 2016, from http://towcenter.org/research/guide-to-automated-journalism/. 31, 32

Hansen, K.A. and Paul, N. (2015a). Editors' comments. Special Issue: "Capturing and Preserving the 'First Draft of History' in the Digital Environment" *News- paper Research Journal* 36(3): 285–287. DOI: 10.1177/07395329156000743. 20, 21

Hansen, K.A. and Paul, N. (2015b). Newspaper archives reveal major gaps in digital age. Special Issue: "Capturing and Preserving the 'First Draft of History' in the Digital Environment" *Newspaper Research Journal* 36(3): 290-298. DOI: 10.1177/07395329156000745. 21

Hare, K. (2015). A new site launches today, offering a place to explore the intersection of news and social media. Retrieved June 26, 2016, from Poynter Institute: http://www.poynter.org/2015/a-new-site-launches-today-offering-a-place- to-explore-the-intersection-of-news-and-social-media/374147/. 23

Hermida, A. (2012). Social Journalism: Exploring How Social Media is Shaping Journalism. In: *Handbooks in Communication and Media*, Volume 38: Handbook of Global Online Journalism. Eugenia Siapera and Andreas Veglis (eds.) Somerset, NJ:John Wiley, pp. 309–328. DOI: 10.1002/9781118313978. ch17. 16, 17, 25

Holt, S. (2016). It's a new career path, not a dead end. *The Bulletin*. March, p. 17. Retrieved May 23, 2016, from *NewsMediaWorks*: http://www.newsmediaworks. com.au/march-bulletin-out-now/. 34

Holzapfel, N. (2016). The future of journalism: Meet my colleague – the robot. 20 January. Retrieved May 2, 2016, from Ludwig Maximilian University, Munich: http://www.en.uni-muenchen.de/news/newsarchiv/2016/graefe_journalism.html. 32, 33

Hooke, P. (2013). Newspapers rise and fall. In: *Challenge and Change: Reassessing Journalism's Global Future*. Alan Knight (ed.) UTS ePress, The University of Technology Sydney, pp. 30–52. 18

Houston, B. (2009). Chapter 1: The Investigative Process. In: *The Investigative Reporter's Handbook: A guide to documents, databases and techniques*. Brant Houston (ed.) and Investigative Reporters and Editors, Inc. Boston: Bedford/St. Martin's. 5th edition, pp. 3–18. 1, 9

Howe, J. (2006). *Crowdsourcing: why the power of the crowd is driving the future of business*. weblog. June 2. Retrieved September 20, 2015, from http://crowdsourcing.typepad.com/cs/2006/06/crowdsourcing_a.html. 15

Ingram, M. (2014). Putting the Human Crowd to Work. In: *Verification Handbook: A Definitive Guide to Verifying Digital Content for Emergency Coverage*, edited by Craig Silverman. Maastricht, the Netherlands: European Journalism Centre. Retrieved February 9, 2016, from http://verificationhandbook.com/book/chapter5.php. 25

Jarvis, J. (2007). For the Record | Comment. *The Guardian*. December 1. Retrieved September 25, 2015, from http://www.theguardian.com/commentisfree/2007/nov/30/fortherecord. 15, 16

Jarvis, J. (2006). "Networked Journalism." *Buzzmachine*. July 5. Retrieved September 25, 2015, from http://www.buzzmachine.com/2006/07/05/networked-journalism.

Jericho, G. (2012). *The Rise of the Fifth Estate: Social Media and Blogging in Australian Politics*. Brunswick, Vic.: Scribe. 16

Journalism that Matters (2011). *Beyond Books: a work session for journalists, librarians and citizens*. April 6–7, 2011. MIT, Cambridge, Mass. Retrieved February 28, 2016, from http://journalismthatmatters.org/biblionews/.

Kovach, B. (2006). Toward a New Journalism with Verification. *Nieman Reports*, Winter, December 15. Retrieved May 7, 2016, from http://niemanreports.org/articles/toward-a-new-journalism-with-verification/.

Kovach, B., and Rosenstiel, T. (2010). *Blur: How to Know What's True in the Age of Information Overload*. New York, NY: Bloomsbury. 15, 16

Library of Congress (2016). Chronicling America: Historic American Newspapers. Retrieved January 29, 2016, from http://chroniclingamerica.loc.gov/.

Los Angeles Times (2016). The Homicide Report. Retrieved April 23, 2016, from http://homicide.latimes.com/.

McCain, E. (2015a). Plans to save born digital news content examined. Special Issue: "Capturing and Preserving the 'First Draft of History' in the Digital Environment" *Newspaper Research Journal* 36(3): 337–347. DOI: 10.1177/07395329156000747. 19

McCain, E. (2015b). Personal correspondence with the author, November 24, 2015. 21

McCain, E. (2015c). Preserving a Visual Record, Part 1. July 23. Retrieved January 29, 2016, from the *JDNA*: https://www.rjionline.org/stories/preserving-a-visual-record-part-1. 19

McCargar, V. (2006). *Missouri J-School and the 'Backstory,'* Retrieved January 29, 2016, from MOspace Institutional Repository: https://mospace.umsystem.edu/xmlui/bitstream/handle/10355/45033/MissouriJSchoolAndTheBackstory2008.pdf?sequence=1. 19

McCosker, A., Reid, D., and Farrell, C. (2016). *Social Media Industries: Bridging the Gap between Theory and Practice*, Department of Media and Communication, Swinburne University of Technology, DOI: 10.4225/50/573D0512BE854. 33

Martin, N. (2014). Information Verification in the Age of Digital Journalism. Presented at the *Special Libraries Association Annual Conference* (Vancouver, B.C., June 8–10) https://www.sla.org/wp-content/uploads/2014/07/Information-Verification.pdf. 2

Martin, N. (2007). Making Intelligent Choices: Career progression between sectors. *InCite* 28(5):8. 11

Martin, N. (2015). Participatory Journalism, Blurred Boundaries: Introducing Theoretical IS Frameworks to Re-orient Research Practices. In: *Providing Quality Digital Information: 17th International Conference on Asia-Pacific Digital Libraries*, ICADL 2015, Seoul, Korea, 9-12 December 2015, Proceedings. LNCS, 9469: 190–196. Springer. DOI: 10.1007/978-3-319-27974-9_19. 3

Matarazzo, J. and Pearlstein, T. (2010). Survival Lessons for Libraries: Staying Afloat in Turbulent Waters. *Searcher* 18(4): 14–16, 18, 46, 48–49, 52–53. 6, 7, 8

Mayer, J. (2011). At the crossroads of journalists and librarians, we find community engagement. *Joy Mayer Journalism+Community*. Retrieved February 28, 2016, from http://joymayer.com/2011/04/07/at-the-crossroads-of-journalists-and-librarians-we-find-community-engagement/. 29

Montgomery, L. (2010). Letter to Craig Silverman, and to the Editor of the *Columbia Journalism Review*, dated February 10. Retrieved May 24, 2016, from the Wayback Machine: https://web.archive.org/web/20150111090003/http://www.cjr.org/behind_the_news/endangered_species.php?page=all. 34

Mumford, W. (2014). Meet your new reporter. *The Bulletin*. Nov 12. Retrieved April 26, 2016, from NewsMediaWorks: http://www.newsmediaworks.com.au/meet-your-new-reporter/. 31, 32

Murdoch, R. (2008). The future of newspapers: moving beyond dead trees. Retrieved November 22, 2015, from Boyer Lectures website: http://www.abc.net.au/rn/boyerlectures/stories/2008/2397940.htm. 17

Murray, A. (2011). BBC processes for verifying social media content. *BBC Academy*, May 18. Retrieved February 9, 2016, from BBC: http://www.bbc.co.uk/blogs/collegeofjournalism/entries/9117dbc6-db68-30fd-9787-41c7da51a85c.

Murray, T.E. (2014). When a Library Shuts Its Doors: Collections and Information Services After a Library Closure, *Journal of Library Administration* 54(2): 147-154, DOI: 10.1080/01930826.2014.903370. 6

Narisetti, R. (2015). Day 2: Plenary – Keynote, paper presented to the *2015 Future Forum: Influencing a Connected World*. Sydney, Australia, September 10–11.

National Public Radio (NPR) (2010). Author interview: 'All Facts Considered' by NPR's long-time librarian. *NPR books*, October 23. Retrieved March 4,

2016, from NPR: http://www.npr.org/2010/10/23/130729448/all-facts-considered-by-nprs-longtime-librarian.

Nelsson, R. (2012). News Libraries. In: *British Librarianship and Information Work 2006-2010*, J.H. Bowman (ed.), Lulu.com, pp. 264–273. 11, 34

Nelsson, R. (2014). Skivers and Strivers and the "National interest": Working with Journalists at the Guardian, Refer: *Journal of the Information Services Group (CILIP)*. Retrieved May 9, 2016, from https://referisg.wordpress.com/tag/news-librarians/. 34

Network Solutions (2016). Whois.net. Retrieved February 5, 2016, from http://www.networksolutions.com/whois/index.jsp.

NTT America (2015). WHOIS behind that domain. Retrieved February 5, 2016, from https://whois.net/.

O'Connor, A. (2007). Special Libraries and Information Services. In: *Libraries in the Twenty-First Century: Charting New Directions in Information Services*, Stuart Ferguson (ed.), Topics in Australiasian Library and Information Studies, Number 27. Wagga Wagga, NSW: Centre for Information Studies, Charles Sturt University, pp. 59–72. DOI: 10.1016/b978-1-876938-43-7.50004-1/. 5, 6

Ojala, M. (2014). Report from the Field. Beyond Borders at SLA. *Information Today*, 31(7): 11–12. 1

Olsson, M. (2014). Information Practices in Contemporary Cosmopolitan Civil Society. *Cosmopolitan Civil Societies Journal* 6(2): 79–93. DOI: 10.5130/ccs.v6i2.3948. 2

One Umbrella Team (2011). Professional Profile. *TOUR: The One Umbrella Report*, 43, 2011. 9

Paul, N. and Hansen, K.A. (2002). Reclaiming News Libraries. *Library Journal* April 1: 44–46. 6

Pew Research Center (2012). The State of the News Media 2012. Retrieved January 29, 2016, from http://www.stateofthemedia.org/2012/mobile-devices-and-news-consumption-some-good-signs-for-journalism/how-community-news-is-faring/. 19

Pew Research Center (2016). The State of the News Media 2016. Retrieved June 21, 2016, from http://www.journalism.org/2016/06/15/state-of-the-news-media-2016/. 15

Phillips, A. (2015). *Journalism in Context: Practice and Theory for the Digital Age*. Routledge, Taylor and Francis Group, London and New York. DOI: 10.1080/21670811.2015.1096622. 24

Poindexter, P. (2015). News Engagement Day should be priority. Special Issue: "Capturing and Preserving the 'First Draft of History' in the Digital Environment" *Newspaper Research Journal* 36(3): 373. DOI: 10.1177/0739532915609141. 18

Posetti, J. (2009). How Journalists are using Twitter in Australia. *Mediashift*. May 27. Retrieved October 1, 2015, from http://mediashift.org/2009/05/how-journalists-are-using-twitter-in-australia147/. 16

Poynter Institute (2016). Poynter - A global leader in journalism. Retrieved June 21, 2016, from http://www.poynter.org/.

ProPublica (2016). ProPublica, Journalism in the Public Interest. Retrieved January 29, 2016, from https://www.propublica.org/.

Quigley, M. (2014). News Library Layoffs and Buyouts. Retrieved October 27, 2015, from https://docs.google.com/document/d/11G5CxLF_lyjzaYu400ATYvr-rlcgTv1uzEgHR-FrWoEY/preview. 6

Reveal (2014). Verify this week (29/2014). Verification becomes a "circular process." Retrieved September 20, 2015, from http://revealproject.eu/verify-this-week-292014/. 2

Riordan, K. (2014). News in the age of digital disruption. *Future Tense – ABC Radio National*. Dec 9. Retrieved April 26, 2016, from http://www.abc.net.au/radionational/programs/futuretense/news-in-the-age-of-digital-disruption/5954106. 35

Rogers, K. (2015). Twitter Cats to the Rescue in Brussels Lockdown. *The New York Times*. Nov 23. Retrieved February 9, 2016, from http://www.nytimes.com/2015/11/24/world/europe/twitter-cats-to-the-rescue-in-brussels-lockdown.html. DOI: 10.1016/s0262-4079(14)60627-8. 24

Rutkin, A. (2014). Rise of robot reporters: when software writes the news. *New Scientist*. 21 March. Retrieved April 30, 2016, from https://www.newscientist.com/article/dn25273-rise-of-robot-reporters-when-software-writes-the-news/. DOI: 10.1016/s0262-4079(14)60627-8. 32

Sabelhaus, L. and Cawley, M. (2013). Searching for News Online: Challenging Traditional Methods. *Online Searcher* 37(2): 10–14. 18, 19

Sanders, C. (2013). A Look to the Future: One Archive Tape at a Time. *NPR Extra*, August 22. Retrieved March 4, 2016, from NPR: http://www.npr.org/sections/npr-extra/2012/08/23/159910472/a-look-to-the-future-one-archive-tape-at-a-time. 28

Schwencke, K. (2014). Earthquake aftershock: 2.7 quake strikes near Westwood. *Los Angeles Times*, March 17. Retrieved April 23, 2016, from http://www.latimes.com/local/lanow/earthquake-27-quake-strikes-near-westwood-california-rdivor-story.html.

Shanley, P. (2015). Breaking News: Robots Invade Newsrooms. April 15. Retrieved April 12, 2016, from LinkedIn: https://www.linkedin.com/pulse/breaking-news-robots-invade-newsrooms-paul-shanley. 32

Shanley, P. (2014). Five Trends That Will Reshape the News Industry. December 9. Retrieved April 18, 2016, from LinkedIn: https://www.linkedin.com/pulse/20141209131730-52209785-5-trends-that-will-reshape-the-news-industry. 35

Shapiro, I., Brin, C., Bédard-Brûlé, I., and Mychajlowycz, K. (2013). Verification as a Strategic Ritual. *Journalism Practice* 7(6): 1–18. DOI: 10.1080/17512786.2013.76538. 1

Shulman, R. (2016). Turning the Page: How the Librarian's Role Continues to Evolve. *Huffington Post*. Retrieved May 17, 2016, from HuffPost Education: http://www.huffingtonpost.com/robyn-shulman-/turning-the-page-how-the-_b_9701694.html. 33

Shumaker, D. (2012). The Embedded Librarian: Innovative strategies for taking knowledge where it's needed, Medford, NJ, *Information Today*, pp. 93–106. DOI:10.1080/0361526X.2013.798851. 6, 7

Silverman, C. (2016). *Regret the Error*. weblog. Retrieved June 26, 2016, from Poynter Institute: http://www.poynter.org/tag/regret-the-error/.

Silverman, C. (2012). A New Age for Truth. *Nieman Reports*, Summer, July 6. Retrieved June 26, 2016, from: http://niemanreports.org/articles/a-new-age-for-truth/. 24

Skinner, K. and Graham, N. (2015). *Scanning the Environment: North Carolina Born-Digital News Preservation Practices 2015*. Educopia Institute, North Carolina Digital Heritage Center: 1–10. 19, 22

Slaughter, A. (2011). Design Your Own Profession. *Harvard Business Review*. December 22. Retrieved May 23, 2016, from https://hbr.org/2011/design-your-own-profession. 34

Sommers, H. (2013). Librarian ProFile: It's Amazing to Hear A Person's Voice Change in their Career. *Inside NPR*, August 26. Retrieved March 4, 2016, from NPR: http://www.npr.org/sections/npr-extra/2013/08/23/214209229/librarian-profile-its-amazing-to-hear-to-a-persons-voice-change-in-their-career. 29

Staff of the Worthington Library (2010). Tracking Trends in the Future of Worthington Library, *Public Library Quarterly* 29(3): 230–271. DOI: 10.1080/01616846.2010.502039. 37

Sterns, J. and Kille, L.W. (2015). Tools for verifying and assessing the validity of social media and user-generated content. *Journalist's Resource*. Harvard Kennedy School Shorenstein Center on Media, Politics, and Public Policy, April 2. Retrieved June 26, 2016, from http://journalistsresource.org/tip-sheets/reporting/tools-verify-assess-validity-social-media-user-generated-content.

Sueiro Bal, M. (2012). NPR Librarian Kee Malesky in New York. *NYPR Archives and Preservation*. Retrieved March 4, 2016, from WNYC: http://www.wnyc.org/story/245470-npr-librarian-kee-malesky-new-york/. 28

Taylor, T.A. and Parrish, J.R. (2009). News Librarian. In: *Career Opportunities in Library and Information Science*. Infobase Publishing, pp. 104–106. 7, 12

Trudell, L. (2014a). Assessing the Credibility of Crowdsourced Content. *Freepint*. 24 October. 36, 37

Trudell, L. (2014b). Personal correspondence with the author, October 3, 2014. 36

University of Maryland (2016). JOUR479M – Understanding Social Media Use in Journalism. Retrieved February 5, 2016, from http://lib.guides.umd.edu/JOUR479M. 25

Vassilakaki, E. and Moniarou-Papaconstantinou, V. (2015). A systematic literature review informing library and information professionals' emerging roles. *New Library World* 116(1/2): 37–66. DOI: 10.1108/NLW-05-2014-0060. 33

Wang, S. (2016). All your insights are belong to us: A new Tow Center report outlines the state of automated journalism. January 7. Retrieved April 11, 2016, from Nieman Lab: http://www.niemanlab.org/2016/01/all-your-insights-are-belong-to-us-a-new-tow-center-report-outlines-the-state-of-automated-journalism/. 31

Wardle, C. (2014). Verifying User-Generated Content. In: *Verification Handbook: A Definitive Guide to Verifying Digital Content for Emergency Coverage*, edited by Craig Silverman. Maastricht, the Netherlands: European Journalism Centre. Retrieved February 9, 2016, from http://verificationhandbook.com/book/chapter3.php. 24

Watkins, J., Park, S., Blood, R.W., Deas, M., Dunne Breen, M., Fisher, C., Fuller, G., Lee, J.Y., Papandrea, F., and Ricketson, M. (2016). *Digital News Report: Australia 2016*. News & Media Research Centre, University of Canberra. DOI: 10.4225/50/5754F7090A5C5. 15

Williams, P. and Nicholas, D. (1997). Journalists, news librarians and the Internet. *New Library World* 98(6): 217–223. DOI: 10.1108/03074809710174306. 5

Zickuhr, K. (2014). Public libraries and technology: From "houses of knowledge" to "houses of access" Pew Internet & American Life Project, July 9. Retrieved October 20, 2015, from http://libraries.pewinternet.org/2014/07/09/public-libraries-and-technology-from-houses-of-knowledge-to-houses-of-access/.

Author Biography

Nora Martin, M.A., B.A., is a Corporate Librarian at the NSW Ministry of Health. She received her Master of Arts in Information and Knowledge Management from the University of Technology Sydney in 2015. Nora was the recipient of the SLA Diversity Leadership Development Program Award (2009) in recognition of implementing the inaugural Library service at AUSTRAC, Australia's anti-money-laundering regulator and financial intelligence unit. Prior to obtaining her M.A., she received a B.A. in Communication (specializing in library science) in 2005. Nora has presented her work internationally, at the 2014 SLA Annual Conference in Vancouver and at the *17th International Conference on Asia-Pacific Digital Libraries*, *ICADL 2015*, held in Seoul, Korea. Her research interests encompass digital humanities, examining the evolving nature of journalists' information practices in the 21st century, and social media research.

Printed in the United States
by Baker & Taylor Publisher Services